SPACE

TRAVEL

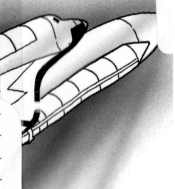

D1303728

Graham

nada Ltd
d Hill, Ontario, Canada

First published in Great Britain in 1991 by
Two-Can Publishing Ltd, 27 Cowper Street, London EC2A 4AP

4321   Printed in Hong Kong   1234/9

Canadian Cataloguing in Publication Data
Graham, Ian, 1953–
Space Travel

Includes index
ISBN 0-590-73876-3

1. Space flight – Juvenile literature.
2. Space vehicles – Juvenile literature. I. Title.

TL795.G73 1991   j629.4   C91-093857-1

Photograph Credits:
p.4 Ann Ronan Picture Library. p.5 (bottom) European Space Agency/Ian Graham. p.5 (right) NASA. p.6. Associated Press. p.7. NASA. p.9 NASA/Spacecharts. p.10 NA
p.11 NASA. p.12 NASA. p.13 NASA/Spacecharts. p.14 NASA. p.15 NASA. p.16 NASA. p.17 NASA/Spacecharts. p.18 NASA/Spacecharts. p19 NASA/Spacecharts.
p.20-21 Julian Baum. p.21 NASA. p.22 Ian Graham. p.30 NASA. p.31 NASA/Spacecharts. Cover photo NASA/Science Photo Library.

Illustration Credits:
All illustrations by Chris Forsey and Peter Bull except those on pages 24 to 28, which are by Virgil Pomfret Artists.

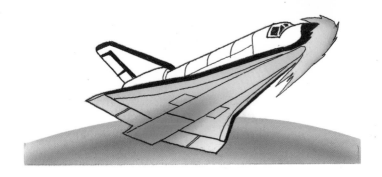

CONTENTS

All words marked in **bold** can be found in the glossary

ROCKETS

Rockets are used to launch spacecraft and **satellites** into space. The first rockets were made in China over 900 years ago. They were propelled by gunpowder and were rather like large fireworks. They were used as military weapons and were not powerful enough to reach space. The technology for making space rockets was not developed until the 1950s.

Modern rockets do not use gunpowder, but are propelled by a variety of **fuels**. Most rocket engines use liquid fuel. This enables them to be turned on and off. Solid fuel rockets cannot be turned off after they are lit. Rockets often have several stages, each with its own rocket engines. When each stage uses up its fuel, its empty fuel tank falls away.

▲ This illustration from a book by Jules Verne, *From the Earth to the Moon*, shows an imaginary space train on its way to the Moon.

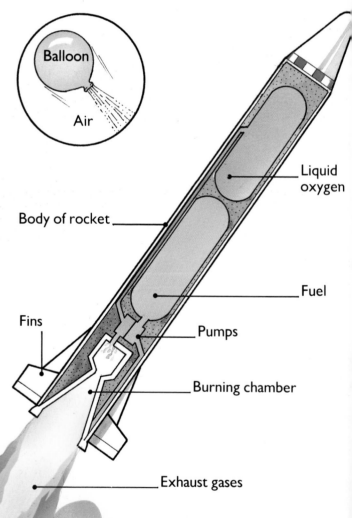

▲ The jet of gases rushing out through a rocket nozzle pushes the rocket in the opposite direction, like a balloon with an open neck.

ROCKET FACTS

The first successful experiments with liquid fuel rockets were carried out by Robert H. Goddard in the United States in the 1920s, as seen in the photo below.

● Rockets are steered by swivelling their nozzles to direct the burning gases in different directions.

● The world's most powerful rocket is the Soviet Energya. It weighs 2,400 tonnes and has a thrust of 4,000 tonnes. It was launched for the first time on 15 May 1987.

● The heaviest object that has ever orbited the Earth was the third stage of the Apollo 15 Saturn 5 rocket. Just before the rocket and spacecraft set off for the Moon, they weighed 140 tonnes.

● A rocket's **thrust** can be increased to launch heavier satellites by strapping extra solid fuel boosters to it.

▲ The Voyager 1 spacecraft is launched on its way to Jupiter and Saturn in 1977 by a Titan 3E-Centaur rocket. This could place over 13 tonnes in Earth orbit or send up to four tonnes to the **planets**.

EARLY SPACEFLIGHTS

Early scientific theories suggested that if an object travelled quickly enough it could resist the force of **gravity**. At a speed of 28,000 kph (17,500 mph), an object should **orbit** the Earth. In 1957 the Soviet Union placed the first artificial satellite, Sputnik 1, in orbit round the Earth.

Four years later the Soviet Union sent the first man, Yuri Gagarin, into space. His Vostok 1 spacecraft was a simple hollow metal ball that made one orbit of the Earth before landing safely. In 1962 John Glenn became the first American to orbit the Earth in a **Mercury** space **capsule**.

DID YOU KNOW?

● The first living creature in space was not a human being, but a Russian dog called Laika launched inside Sputnik 2 on 3 November 1957. Laika died in space.

● In August 1960, two dogs, Belka and Strelka, were launched into space by the Soviet Union. They were brought back safely to Earth after 24 hours.

● During John Glenn's Mercury flight, an indicator showed that the capsule's heat shield was not locked in position. Without it, the capsule would burn up when it re-entered the **atmosphere**. In fact, the indicator was faulty and the capsule re-entered safely.

▲ Astronaut John Glenn is prepared for a test before the United States' first manned orbital spaceflight in 1962. Glenn took five hours to make three Earth-orbits.

◀ The world's first spaceman, the Soviet cosmonaut Yuri Gagarin, waits to board his Vostok 1 spacecraft.

▶ US astronaut Thomas Stafford (far right) visits Soviet cosmonaut Valery Kubasov during the Apollo Soyuz Test Project in 1975. The US Apollo and Soviet Soyuz spacecraft linked up in Earth orbit.

SATURN ROCKET

The Saturn 5 rocket was designed to take US **astronauts** to the Moon. It was built in three stages to lift the 53-tonne Apollo spacecraft into Earth orbit and then put it on course for the Moon.

The first two stages were dropped into the Atlantic Ocean when their fuel was used up. The third stage was fired twice – once to place the spacecraft in Earth orbit and a second time to push it out of orbit towards the Moon. When this was achieved it was sent into orbit around the Sun or made to crash on to a carefully selected part of the Moon to help with scientific studies of the Moon's structure.

▶ A Saturn rocket with an Apollo spacecraft on top stood 111 m (363 feet) high on the launching pad. In seven years, these giant rockets launched seven Apollo crews and placed one Skylab **space station** in Earth orbit.

DID YOU KNOW?

● The Saturn 5's first-stage rocket engines were the most powerful ever built. They burned 13 tonnes of fuel per second, producing 3,400 tonnes of thrust to lift 3,000 tonnes off the launching pad.

● Fifteen Saturn 5 rockets were built at a cost of $6 billion US.

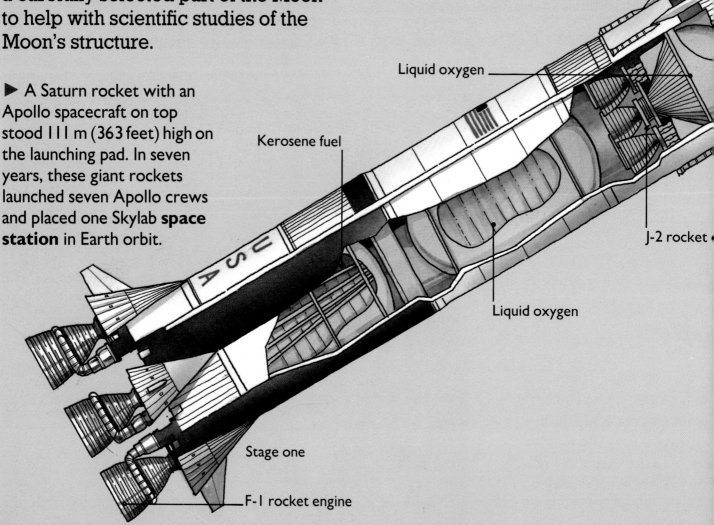

Stage two

Liquid oxygen

Kerosene fuel

J-2 rocket

Liquid oxygen

Stage one

F-1 rocket engine

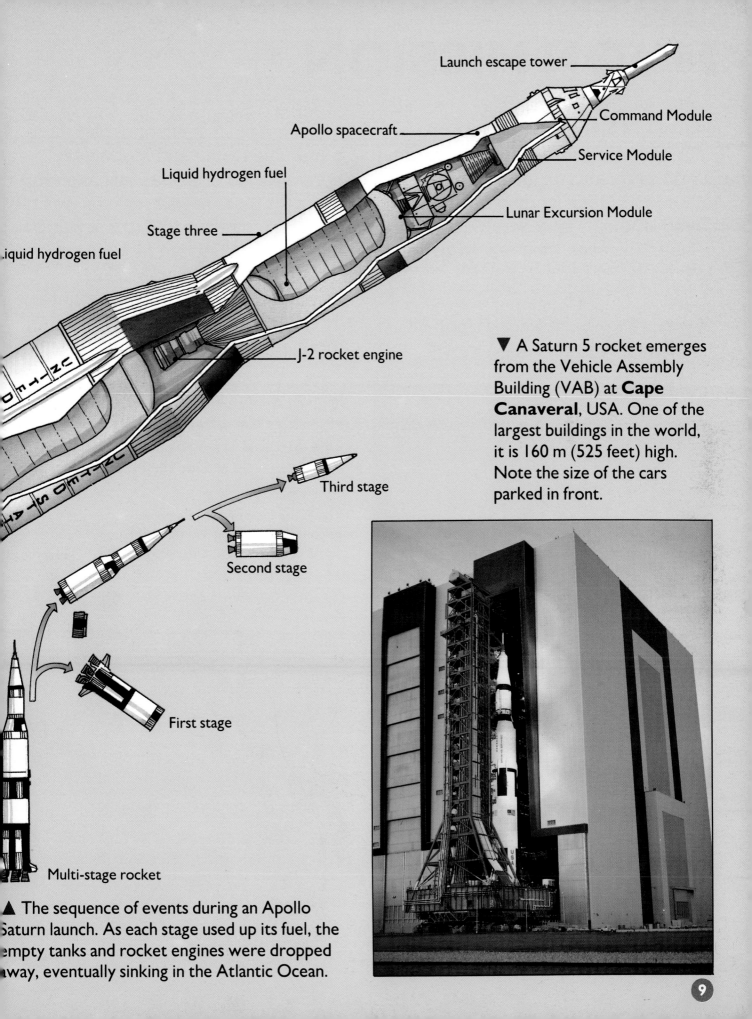

Launch escape tower

Command Module

Apollo spacecraft

Service Module

Liquid hydrogen fuel

Lunar Excursion Module

Stage three

Liquid hydrogen fuel

J-2 rocket engine

Third stage

Second stage

First stage

Multi-stage rocket

▲ The sequence of events during an Apollo Saturn launch. As each stage used up its fuel, the empty tanks and rocket engines were dropped away, eventually sinking in the Atlantic Ocean.

▼ A Saturn 5 rocket emerges from the Vehicle Assembly Building (VAB) at **Cape Canaveral**, USA. One of the largest buildings in the world, it is 160 m (525 feet) high. Note the size of the cars parked in front.

MOON MISSION

Between 1969 and 1972 six Apollo missions were responsible for landing 12 US astronauts on the Moon. The attempted Moon landing of Apollo 13 failed when the spacecraft developed a fault. However, the crew were able to land safely back on Earth.

▼ A Saturn 5 rocket prepared for blast-off from its launching pad at Cape Canaveral.

Each Apollo spacecraft had three parts. The Command Module, a cone-shaped craft with just enough room for the three astronauts, was the only part to return to Earth. It was attached to a Service Module until just before the Command Module re-entered the Earth's atmosphere. The astronauts descended to the Moon in a Lunar Excursion Module.

▼ Two Apollo astronauts would land on the Moon's surface in the Lunar Excursion Module. One astronaut stayed behind in the Command Module in orbit around the Moon.

The Lunar Module rockets blasted it from the launching pad to the Command Module circling above.

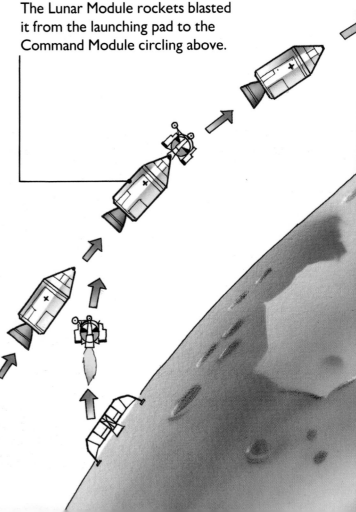

MOON FACTS

● The first people to travel beyond Earth orbit were the three crew members of Apollo 8 who circled the Moon in 1968. The 400,000 km (250,000 miles) journey took three days.

● By the time Apollo 17 astronauts Eugene Cernan and Harrison Schmitt left the Moon for the last time in 1972, the Apollo Project had cost $25 billion.

▼ The Apollo Moon explorers blasted off from the Moon to rejoin the Command Module above. Once inside the Command Module, they began their return journey to Earth.

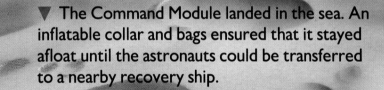

The Command Module headed back home for re-entry through the Earth's atmosphere.

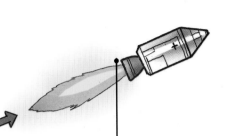

▼ The Command Module landed in the sea. An inflatable collar and bags ensured that it stayed afloat until the astronauts could be transferred to a nearby recovery ship.

MAN ON THE MOON

During their short stays of 1-3 days on the Moon, Apollo astronauts carried out many scientific experiments and collected Moon rocks for analysis on Earth. Some experiments were left on the Moon to continue working after the astronauts had returned to Earth.

The Apollo 11, 12 and 14 crews had to move around the Moon on foot. With spacesuits on, they could walk 4 km (2.5 miles) at most. Apollos 15, 16 and 17 carried a battery-powered car called a Lunar Rover which enabled astronauts to travel much further.

▲ Apollo 11 astronaut Edwin "Buzz" Aldrin steps on to the Moon's surface. The backpack enabled the astronaut to work for several hours outside the spacecraft.

◀ The Lunar Rover – a car designed specially for driving on the Moon's dusty surface.

Before going on to the Moon's airless surface, the astronauts had to put on spacesuits and backpacks. The backpacks provided an air supply for them to breathe. They also stopped the astronauts from getting too hot by pumping cool water through tubes in the spacesuits. Each was fitted with a radio to keep the astronauts in contact with each other and with mission controllers on Earth.

The astronauts also used radios to send information about **moonquakes** and the Moon's magnetic field to Earth.

DID YOU KNOW?

● The Lunar Rover weighed 209 kg (460 pounds) and could travel at up to 16 kph (10 mph) on level ground. Three Lunar Rovers were left on the Moon.

● The Apollo 17 astronauts Eugene Cernan and Harrison Schmitt travelled the furthest distance around the Moon's surface, a total of 34 km (21 miles).

● The six Apollo crews that landed on the Moon brought a total of 379 kg (836 pounds) of Moon rock and soil back to Earth with them. The Apollo 17 crew alone collected 110 kg (243 pounds).

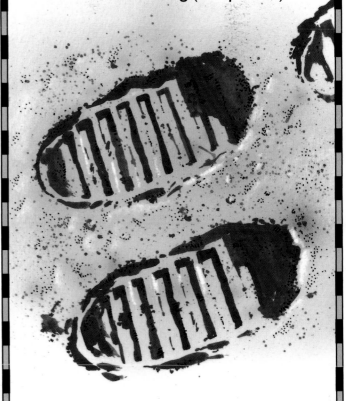

● The first footprints on the Moon were made by Neil Armstrong on 20 July 1969. They will stay there unchanged as there are no winds to blow them away and no rain to wash them out.

SPACE SHUTTLE

Until the first Space Shuttle mission by the United States in 1981, all rockets and spacecraft could be used only once. The Space Shuttle was designed to be used again and again to reduce the cost of spaceflights.

At liftoff the Space Shuttle weighs about 2,000 tonnes. The Orbiter is the only part that travels into space and returns to Earth. It has a big **payload** bay for carrying satellites into space. During **re-entry**, air rubbing against the Orbiter heats parts of its surface to over 1,500°C (2,700°F).

▶ The US Space Shuttle Orbiter with its payload bay doors open. The Orbiter's robot arm is being used to move a satellite out of the bay and place it in orbit round the Earth.

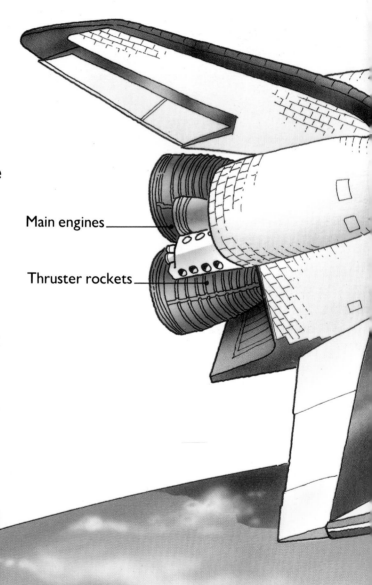

Main engines

Thruster rockets

SHUTTLE FACTS

● The Orbiter has 44 small rocket engines in its nose and tail, used to make small adjustments to its position.

● At takeoff, the Space Shuttle's three main engines use up the 2 million litres (almost 500,000 gallons) of fuel in its external fuel tank in only 8.5 minutes. Two booster rockets provide 80 per cent of the thrust needed for takeoff.

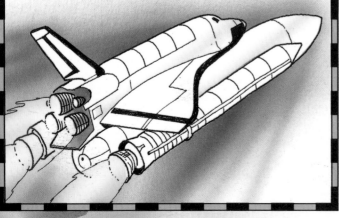

▲ The Space Shuttle Orbiter resembles an airplane. Its wings and tail allow it to turn in flight within the Earth's atmosphere. It glides in to land more steeply than a passenger jet.

Satellite

Robot arm

Payload bay

Flight deck

Heat-resistant tiles

Payload bay door

▶ With all engines firing, the US Space Shuttle blasts off for space. All the main parts of the Space Shuttle – the Orbiter, the main fuel tank and the two booster rockets – can be seen.

SHUTTLE MISSIONS

The Space Shuttle was designed mainly as a "space truck" to carry satellites – communications, spy or scientific research – into orbit.

In space, the Orbiter's payload bay doors open and the satellite is released into space by the Orbiter's 15 m (50 foot) long robot arm or by springs. At a safe distance from the Orbiter, a rocket motor attached to the satellite then fires and boosts the satellite into its correct orbit.

There is enough room inside an Orbiter for seven astronauts. Shuttle missions may last for up to 30 days.

DID YOU KNOW?

● 32,000 tiles glued on the underside of the Orbiter act as a vital heat shield during re-entry.

● A laboratory called Spacelab can be carried in the Orbiter's payload bay.

▼ Mission over, a Space Shuttle Orbiter lands at over 300 kph (190 mph) on the dry lake bed at the Edwards Air Force Base in California.

▶ Astronauts check equipment in the Orbiter's payload bay. A tether or lifeline stops them floating away.

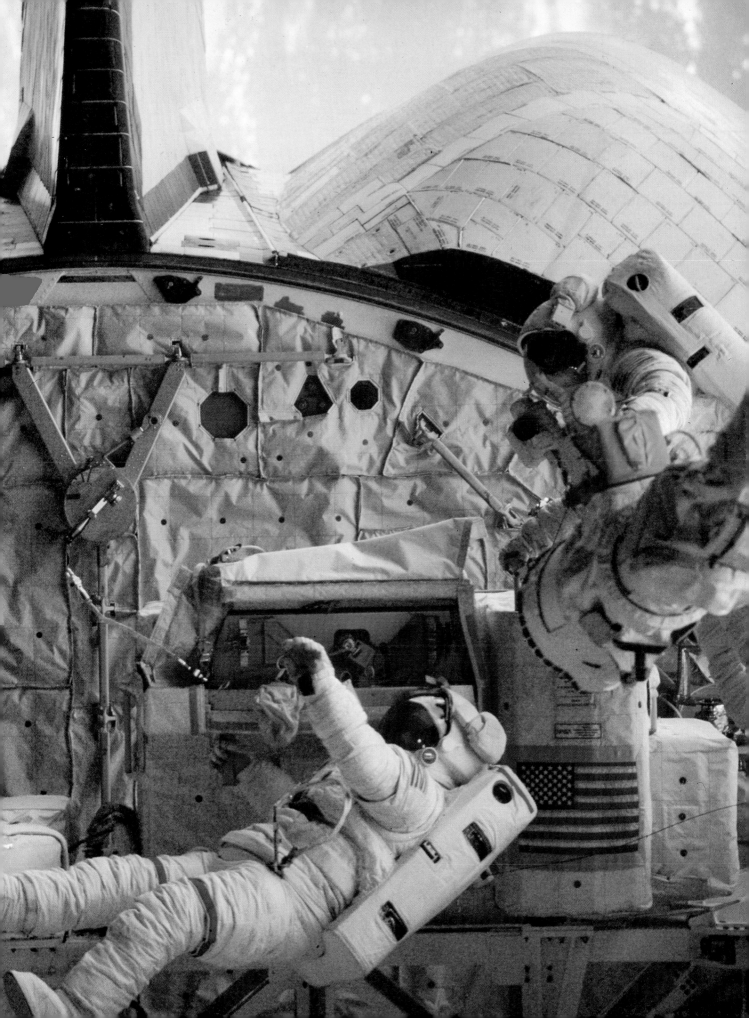

SPACE WALKING

Future astronauts will sometimes have to leave their spacecraft to build and repair structures such as space stations. This is called spacewalking or extravehicular activity (EVA).

Astronauts started practising this in the 1960s. The Soviet **cosmonaut** Alexei Leonov became the first spacewalker on 18 March 1965 during the Voshkod 2 mission. He was tied to his craft by a tether. The first astronaut to leave his craft without a tether was Bruce McCandless on 7 February 1984. He used a Manned Manoeuvring Unit, or MMU, propelled by gas jets.

▶ This artist's impression of a future space station shows a Space Shuttle Orbiter docking with the station, perhaps to deliver fresh supplies.

▼ Astronauts building and inspecting the space station will fly around using personal spacecraft similar to the Space Shuttle's Manned Manoeuvring Unit.

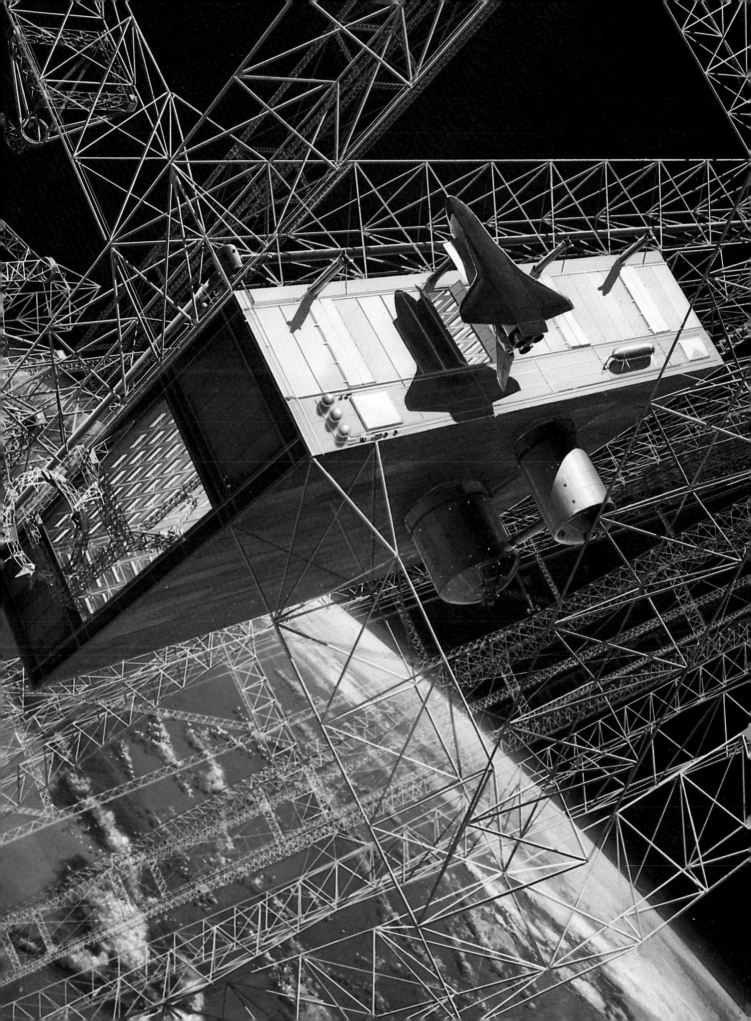

MAN ON MARS

The Soviet Union and the United States are both planning manned missions to Mars. A Mars mission would be much more difficult than the Apollo Moon missions. Mars is over 54 million kilometres (34 million miles) from Earth, over 140 times farther than the Moon. It takes nine days to fly to the Moon and back. It would take around 900 days to fly to Mars and back.

Supplies necessary for such a flight would weigh at least 15 tonnes. The spacecraft would be so large that it would have to be built in Earth orbit. Weightlessness during such a long mission could have undesirable effects on astronauts, including loss of calcium from their bones. Soviet cosmonauts have spent up to 237 days in the *Mir* space station studying this.

An impression of a spacecraft in orbit around Mars with two astronauts spacewalking. The craft is fitted with ports where Mars landing modules can dock.

This is how a future base on the Martian surface may look. Two astronauts wearing spacesuits are leaving their living quarters, while a spaceplane glides in to land.

MARS FACTS

The most advanced spacecraft to reach Mars so far were two US unmanned Viking craft. In orbit round Mars, each split in two. The Lander descended to the surface, as shown below, while the Orbiter circled the planet.

Vikings 1 and 2 studied the Martian atmosphere and analyzed soil samples for signs of life.

SPACE STATION

A space station is a large spacecraft designed to stay permanently in orbit. Crews live and work in space stations for several months at a time. The Soviet Union launched the first space station, which was called Salyut 1, in 1971. Two years later, the United States launched its own small space station, Skylab.

In 1987, the Soviets launched a newer and bigger space station called *Mir*. The United States is now planning to build an international space station where scientists from different countries will work together.

A Space Shuttle docks with a space station. The plates on the sides of the station are solar panels that make electricity from sunlight.

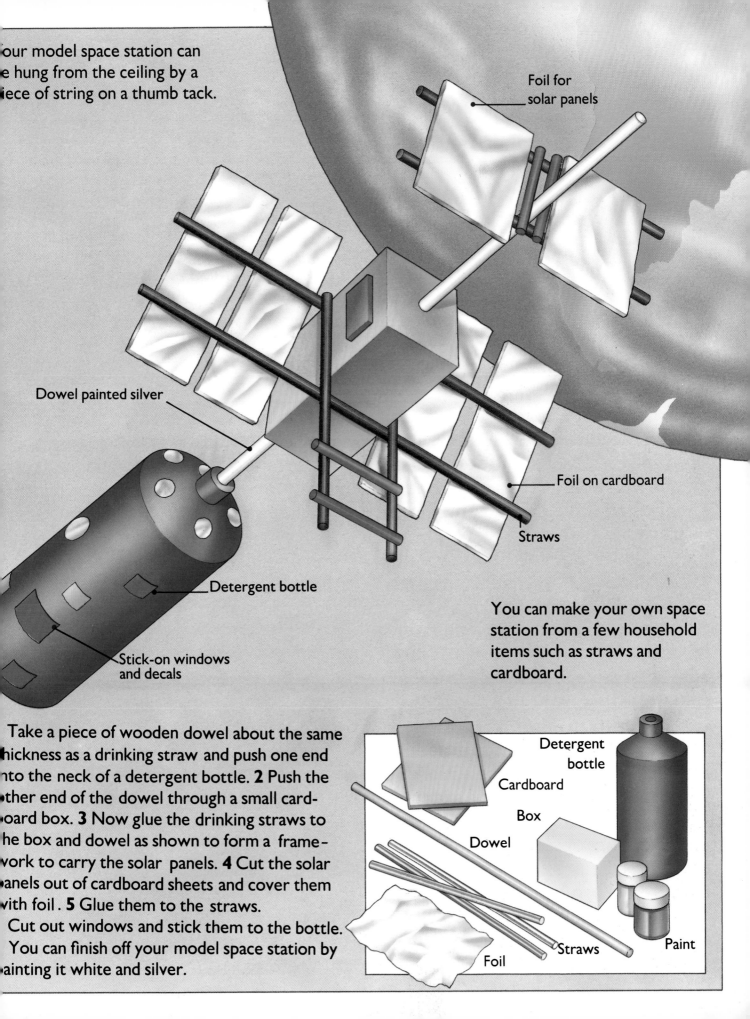

Your model space station can be hung from the ceiling by a piece of string on a thumb tack.

Foil for solar panels

Dowel painted silver

Detergent bottle

Stick-on windows and decals

Foil on cardboard

Straws

You can make your own space station from a few household items such as straws and cardboard.

Take a piece of wooden dowel about the same thickness as a drinking straw and push one end into the neck of a detergent bottle. 2 Push the other end of the dowel through a small cardboard box. 3 Now glue the drinking straws to the box and dowel as shown to form a framework to carry the solar panels. 4 Cut the solar panels out of cardboard sheets and cover them with foil. 5 Glue them to the straws.

Cut out windows and stick them to the bottle. You can finish off your model space station by painting it white and silver.

Detergent bottle

Cardboard

Box

Dowel

Foil

Straws

Paint

The Stowaways

This is an imaginary story but it is based on fact. The year is 2150, and the place is the Harding Space Center in San Antonio, Texas. From here spaceplanes blast off every week for space stations in Earth orbit. This is the story of two children determined to visit their friends in space.

Peter and Sue crouched down at the back of the small blue bus and hid. They felt the bus wobble as someone climbed on board. The engine started and the bus moved off. After a short drive it came to a halt. Peter and Sue peeped over the seats and saw the driver and two passengers leaving. They could also see the Clipper 3 spaceplane towering above them. The passengers were the two astronauts who would fly the Clipper 3 to a space station located many thousands of kilometres from Earth.

Peter and Sue left the bus unseen. An elevator that had taken the astronauts to the top of the spaceplane's launch tower descended to ground level again and the doors opened. There was no one inside. Peter and Sue walked in and pressed the top button marked 'LEVEL 4'. When the elevator doors opened again, they found themselves at the top of the tower.

A small room linked the elevator to a hatch in the side of the spacecraft. The children quickly stepped through the hatch. Above their heads they could see the two astronauts lying on their backs in seats on the flight deck. Peter and Sue clambered down to the mid-deck below the flight deck. The

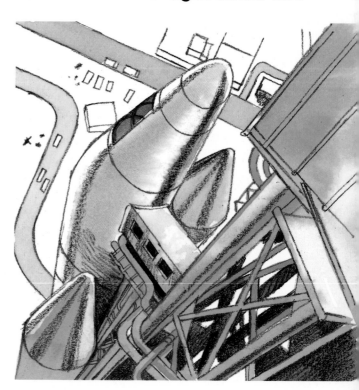

ceiling here was covered with storage bins. Underneath them were two spare seats. Peter sat in one of them and fastened the seat belt.

Suddenly, they heard a rumbling noise and the spacecraft began to move to and fro. Sue quickly strapped herself into the other seat. The rumbling changed to a roar. They could feel themselves being pressed back into their seats. Clipper 3 was taking off!

In the Mission Control centre, scientists and engineers monitoring the mission noticed several strange things. A dotted line on a computer screen showed the flight path predicted for the craft. A second line showed its actual flight path. The two should have been exactly on top of one another, but they were not. The spacecraft was lower than predicted. Puzzled, Mission Control advised Clipper 3's commander to increase engine power. It was as if the spacecraft was a fraction heavier than it should have been. They rechecked their readings.

On another screen, the life support system was being monitored. It showed that the air supply was being used up more rapidly than expected. Perhaps there was an air leak. An engineer checked the cabin pressure and all main valves on her computer screen. Everything seemed correct. The only explanation she could think of was that a living creature or creatures on board the spacecraft were adding to its weight and breathing the air.

A few minutes later, Sue saw the pen from Peter's shirt pocket float away to the other side of the cabin. She tried to catch it, but it was too far away. Unfastening her belt, she reached out to the pen and found herself floating off her seat. They must be in orbit around the Earth, she realized. Peter caught hold of her arm and pulled her back down ... except that there seemed to be no up and down any more. Whichever way Sue floated seemed to be the right way up.

They heard a voice saying, "That's confirmed, Mission Control, we have two extra passengers." The children looked towards the flight deck and saw an astronaut at the radio controls. They could see that he was angry – and no doubt, with them.

"Well at least you had the good sense to strap yourselves in properly for launch. Don't you realize you could have been hurt?" he asked.

The children nodded. "We just wanted to see our friends on *Orion*," said Sue. *Orion* was one of the three US research stations in orbit round the Earth.

"You are going to *Orion*, aren't you?" Peter asked anxiously.

"We are, but you're going home," said the pilot angrily. "Our fuel and supplies are carefully calculated so that we carry no excess weight. If you stay with us all the way, we'll have no margin for safety on our way back. Fortunately, there's another Clipper leaving *Orion* today. We've arranged

for it to take on extra fuel and meet up with us so that you can be transferred across to it."

He could see that they were very disappointed. "Well," he said, in a more kindly voice. "As long as you're here, you might as well come up front and enjoy the view."

He guided them to the flight deck. The commander was sitting in his seat on the left. The pilot strapped Sue into the seat on the right and floated between the two seats with Peter.

Panels in front of them, above their heads and along the sides of the flight deck were covered with hundreds of switches. Three computer screens showed information about the craft. Outside the window, they could see the Earth slowly turning beneath them. The blue ocean and brown land were covered by fluffy white clouds. There was no sound apart from the hissing of the air-conditioning system.

The commander pointed out a tiny speck of light. It looked like a star, and it was growing bigger. It was Clipper 5. The two craft manoeuvred close to one another. The children were transferred from one craft to the other inside a pressurised module that looked like a large soccer ball with a window. A tether attached the module to the pilot's jet-propelled backpack.

When Peter and Sue were safely strapped into their seats inside Clipper 5, it started back to Earth. The children could feel the spacecraft vibrate and bounce as it flew through the thinnest air at the top of the atmosphere. Then they began to feel heavier. A red glow appeared around the windows as the spacecraft was heated red-hot by the air outside rubbing against it. The silence of space was gradually replaced by the sound of air rushing past outside.

They were soon flying through the atmosphere like an airplane. A few minutes later, there was a bump as the wheels touched down on the runway a the Space Center and the spacecraft taxied to a halt. A small blue bus like th one in which they started their adventure was waiting to drive them back to Mission Control.

TRUE OR FALSE?

Which of these facts are true and which ones are false? If you have read this book carefully you will know the answers.

1. The first man in space was the Soviet cosmonaut Herman Titov.

2. Mercury space capsules carried the first Americans into space.

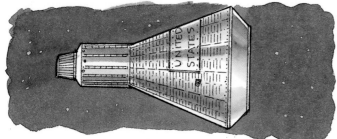

3. The most powerful rocket in the world, Saturn 5, launched Soviet space stations.

4. To orbit the Earth a satellite must reach at least 4,000 kph (2,500 mph).

5. Each successful Apollo mission landed two US astronauts on the Moon.

6. The crew members of Apollo 14 drove around the Moon in a Lunar Rover.

7. The Space Shuttle is designed to be used again and again to carry satellites into orbit.

8. When an astronaut leaves his or her spacecraft and steps out into space, this is called a space walk or extravehicular activity.

9. The first person to go on a space walk was Alexei Leonov in 1965.

10. The two Voyager spacecraft were sent to Mars by the United States just to look for signs of life there.

11. *Mir* is the name of a Soviet space station.

12. Astronauts had to wear spacesuits on the Moon to prevent them from sneezing because of all the moondust in the air there.

ANSWERS: 1.False; 2.True; 3.False; 4.False; 5.True; 6.False; 7.True; 8.True; 9.True; 10.False; 11.True; 12.False.

29

GLOSSARY

● **Astronauts** are people trained to travel into space. Astronaut means star traveller.

● **Atmosphere** is the layer of gas surrounding a planet. The Earth's atmosphere consists mostly of nitrogen but has lots of oxygen.

● **Cape Canaveral** is the place on the Florida coast in the United States where most US rockets are launched.

● **Capsule** is the name given to a small spacecraft designed to travel into space and obtain scientific information.

● **Cosmonaut** is the Soviet equivalent of an astronaut.

● **Fuels** are materials which can be burned to produce energy.

● **Gravity** is the force that attracts objects towards each other.

● **Mercury** is the name of the first series of US manned spacecraft. Each Mercury capsule carried one person. Mercury is also a planet.

● **Moonquakes** are like earthquakes except that they happen on the Moon.

▲ The Space Shuttle Orbiter *Challenger*, with its payload bay doors open, circles the Earth above the atmosphere.

● **Orbit** is the curved path followed by an object circling a planet or star.

● **Payload** is the name for the cargo, usually satellites or other scientific equipment, carried by a rocket or Space Shuttle.

● **Planets** are objects that revolve around a star. The planets that revolve around our Sun are Mercury, Venus, Earth, Mars, Jupiter, Saturn, Uranus, Neptune and Pluto.

● **Re-entry** describes the return of a spacecraft to Earth through the

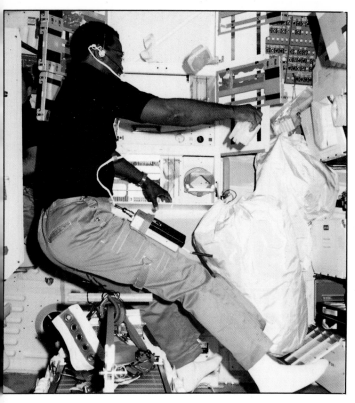

▲ Astronaut Ronald E. McNair, weightless in space, prepares a meal for his fellow Shuttle-crew members.

atmosphere. The re-entry angle must be correct or the craft will either burn up or bounce back into space.

● **Satellite** is an object that revolves around a planet. It might be a natural satellite, such as the Earth's Moon, or a man-made satellite, like Sputnik 1.

● **Space Station** is a large structure permanently in orbit around a planet, in which people live and work.

● **Thrust** is the force that makes a rocket move. It is produced by burning fuel in a rocket engine.

SPACE TIMECHART

1957 The Soviet Union launches the first man-made object into space – Sputnik 1.

1959 Luna 1 is the first man-made object to escape the Earth's gravity on its way to the Moon.

1961 Soviet Yuri Gagarin is the first spaceman.

1962 John Glenn becomes the first US astronaut to orbit Earth.

1962 The first live TV pictures are relayed across the Atlantic Ocean by the Telstar satellite.

1969 Apollo 11 Lunar Module makes the first manned landing on the Moon.

1976 Vikings 1 and 2 land on Mars.

1979 The Voyagers photograph Jupiter.

1981 First US Space Shuttle flight.

1986 A Space Shuttle explodes just after takeoff. The *Challenger* Orbiter is destroyed killing the crew of seven.

1988 The first Soviet Space Shuttle is launched into space, unmanned.

1989 Voyager 2 photographs Neptune.

INDEX